Antarctic Animals

Melvin and Gilda Berger

SCHOLASTIC INC.
New York Toronto London Auckland Sydney
Mexico City New Delhi Hong Kong Buenos Aires

Photographs: Cover: Fritz Polking/Dembinsky Photo Assoc. (DPA) p. 1: Rod Planck/DPA ;
p. 3: Mark J. Thomas/DPA; p. 4: Bryan & Cherry Alexander/arcticphoto.co.uk;
p. 5: Johnny Johnson/Stone/Getty Images; p. 6: John Shaw/Bruce Coleman Inc.;
p. 7: Des Bartlett/Bruce Coleman Inc.; p. 8: John Hyde/Bruce Coleman Inc.;
p. 9: Marilyn Kazmers/DPA; p. 10: John Shaw/Bruce Coleman Inc.;
p. 11: Norbert Wu/Minden Pictures; p. 12: Rod Planck/DPA;
p. 13: Frans Lanting/Minden Pictures; p. 14: Rod Planck/DPA ;
p. 15: Erwin & Peggy Bauer/Bruce Coleman Inc.; p. 16: Fritz Polking/DPA.

Photo Research: Sarah Longacre

ISBN 0-439-67900-1

12 11 10 9 8 7 6 5 4 6 7 8 9 10/0

Printed in the U.S.A.
First Scholastic printing, January 2005

Antarctic animals live around the South Pole.

Penguins live in the Antarctic.

They swim in the sea.

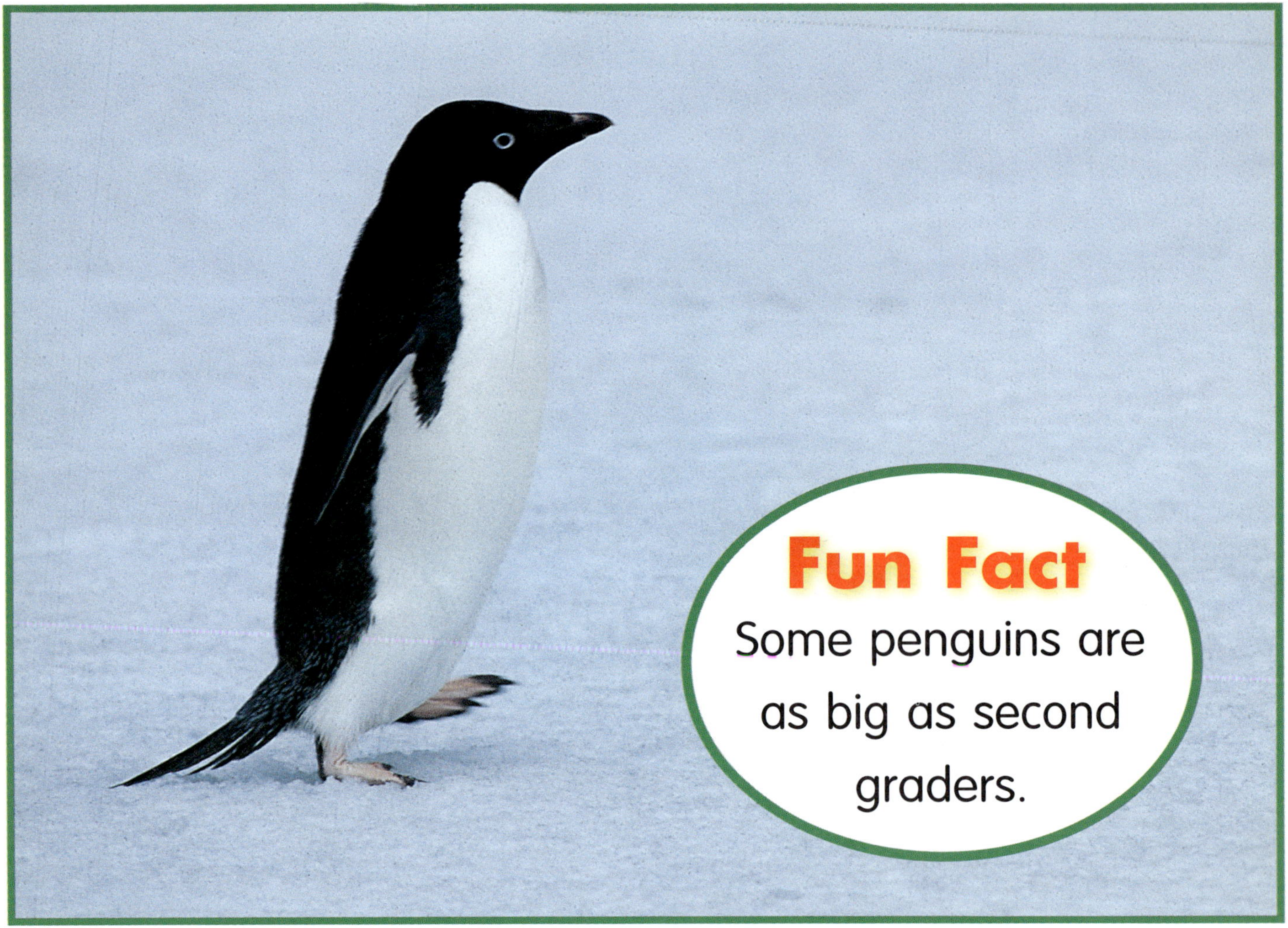

Fun Fact

Some penguins are as big as second graders.

Penguins walk on ice.

They lay eggs on land.

Whales live in the Antarctic.

They breathe air.

Seals live in the Antarctic.

They swim fast.

Skuas (SKOO-uhz) live
in the Antarctic.

They eat penguin eggs.

Albatrosses (AL-buh-traw-suhs) live in the Antarctic.

They eat fish from the sea.

Would you like to visit
the Antarctic?